Mental Math

Shortcut Tricks to Improve Number Manipulation Skills 1

David Itanola

b

No part of this publication may be reproduced, or transmitted in any form or by any means including photocopying, recording or other electronics or mechanical method without the prior permission of the publisher except for brief quotations embodied in critical reviews.

Copyright © 2018 David Itanola

ISBN-13:
978-1986747882

ISBN-10:
1986747883

Table of contents

An overview of this book iii

Introduction v

1. A Quick Guide to Mental Math Strategy 1
2. Divisibility Rules 7
3. Shortcut Tricks for Multiplication and Division of Numbers 18
4. Shortcut Method Multiplication of a Number by 11 46
5. Shortcut Strategies for Calculating Squares of Numbers 56
6. Shortcut Tricks for Square Roots of Numbers 74
7. Shortcut Strategies for Calculating Cube roots Of Numbers 80
8. Decimal Fractions 85
9. Finding the Units Digits of Large Powers 92
About the Author 106

An Overview of this Book

This insightful book teaches how to do mental math with techniques designed to simplify basic math operation for fast and easy number manipulation. Problem-solving procedures are stated systematically in step by step for easy understanding. Examples and problems are completed with answers for practice and improving your memory and computation ability. The book contains relevant tables and diagrams with facts presented clearly in an organized fashion to aid quick understanding of concepts. All you need to do is study the chart and jump to the answer. The book could help the reader in countless ways depending on his needs and scope of practice. It could be a valuable resource for anyone preparing for standardized tests and various competitive examinations. Its simple language and orderly presentation of facts make the book a delight of everyone.

By reading this book, you will be able to:

• Solve problems using multiple strategies that reinforce number sense, which can be helpful in competitive examinations.

- Learn quicker methods by observing some simple techniques.

- Save some precious minutes in various competitive and standardized tests by employing the tricks and rapid calculation techniques.

- Understand the underline tricks to quick number manipulation thereby develop personal tools in the field of quick calculations.

Introduction

Math is everywhere; we come across math every day. Number manipulation can be a challenging task, but the good thing is that math problem can be solved in a variety of ways depending on the level of understanding of the learner. Once you understand the underlying tricks to quick manipulation, you can do it in your own way and develop a trusted efficient and flexible method. Even if you are a non-figure person and scare of math, this book can change your orientation and help you to become an instant math genius. You can use the simple techniques in this book to do math faster than a calculator effortlessly in your head, even if you have no inclination for math.

All you need to do is to practice the techniques to get comfortable using them. It may take some time and energy, but the efforts will pay off in the long run.

1. A Quick Guide to Mental Math Strategies

Understanding Key Terms in Math Operation is crucial to success in number fluency and mental math. This chapter is devoted to the explanation of some common basic terms relating to operation with numbers. The aim is to make the book useful to everyone including those with little knowledge of math. If you are a *non-figure* person, I will advise you memorize them if it is possible or at the least, you should have them written down for quick reference.

If you are already familiar with the terms, having them revised will refresh your memory and prepare you for a journey into the fantastic world of numbers. Above all, understanding these basic vocabularies will be your quick guide towards greater metal math success.

Addition

Addends: The numbers being added together.

Sum: The answer to the addition of two numbers. The sum of 5 and 8 is 13. **5+8=13**

Subtraction: The difference between two numbers

Consider this equation   **18-14=4**

The number **18** is referred to as minuend, **14** is the subtrahend while **4** is the difference.

Multiplication

Factors: The number being multiplied.

Product: The answer to multiplication problem.

Multiplier: One of the factors of the products of two numbers

The product of 5 and 4 is 20.   **5x4=20**

Division

Dividend: The number you need to divide.

Divisor: The number you will use to divide.

Quotient: The answer to the division problem.

Remainder: A leftover when the dividend could not be divided evenly.

Suppose we divide 97 by 6 i.e. $97 \div 6$

```
                  1 6      → Quotient
Divisor ← 6 | 97          → Dividend
              96
              ---
               1          → Remainder
```

This can be illustrated as:

Dividend = (Divisor x Quotient) + Remainder

Or

Divisor = [(Dividend)-(Remainder)] / Quotient

Properties of Numbers

Commutative property

Any finite sum or product of two numbers is unchanged by re-ordering its terms or factors. When you add or multiply two numbers, the arrangement of the number does not affect the outcome.

For example

a+b=b+a

5+4=9 is the same as 4+5=9

Similarly

axb=bxa

5x4=20 is the same as 4x5=20

Additive Identity

If you add zero to a real number or add a real number with zero then you have the same real number. 0+5=5 or 5+0=5. '**0**' is thus referred to as additive identity.

Multiplicative Identity

The number 1 is the multiplicative identity. The product of any number with 1 leaves the number unchanged.

1x5=5x1=5

Additive Inverse

When you add two numbers together and the result gives you '0' the additive identity; then such numbers are inverses to each other. In the set of natural counting numbers; 1 is the inverse of -1 because when you add the two numbers you get 0.

1+ (-1) =0

-5+5=0

Multiplicative Inverse

A multiplicative inverse or reciprocal for a number x, denoted by 1/x or x^{-1}, is a number which when multiplied by x yields the multiplicative identity, 1

For example

1. 2 and ½;

2. 2/3 and 3/2

1. 2 x ½=2/2=1

2. 2/3 x 3/2 =6/6=1

Distributive law

Distributive law is the **law** connecting the operations of multiplication and addition in math, stated symbolically thus, a(b + c) = ab + ac; that is, the monomial factor a is distributed, or separately applied, to each term of the binomial factor b + c, resulting in the product ab + ac.

Or simply stated as multiplying a number by a group of numbers added together is the same as doing each multiplication one by one.

Example: 5 × (3 + 8) = 5×3+ 5×8

The terms in the LHS is the same with the terms in the RHS

Number Classification

Natural numbers: All counting numbers without zero (1, 2,3,4,5....∞)

Whole numbers: Natural numbers with zero (0, 1, 2,3,4,5...∞)

Integers: All whole numbers including negative and positive numbers (......-4,-3,-2,-1, 0, 1, 2, 3, 4,5....∞)

Even numbers: All whole numbers divisible by 2 (2, 4, 6, 8, 10...)

Odd numbers: All whole numbers not divisible by 2 (1,3,5,7,9,11,13,15,17,19....∞)

Prime numbers: A number that has two factors itself and 1 (2,3,5,7,11,13,17,19,23,29,31,37,41,43,47,53,59,61..

Composite numbers: Natural numbers which are not primes

Co-prime: Two natural number x and y are said to be co-prime if their HCF is 1.

2. Divisibility Rules

Divisibility rules are efficient shortcut methods to check whether a given number is evenly divisible by another number. What it means is that when you divide a number with another number, the result is a whole number i.e. no remainder or leftovers.

This is a good shortcut trick in mental math as it helps in determining whether a given integer is divisible by a fixed divisor by examining the digits without necessarily performing the operation. We shall examine divisibility rules for 1-19.

Divisibility by 2:

Rule: A number is divisible by 2 if it ends with 0,2,4,6 or 8

Example 50, 422, 524, 9158, 242 are all divisible by 2 but 2345, 19607 are not

Divisibility by 3:

Rule: A number is divisible by 3 if the sum of it digits is divisible by 3 or multiples of 3.

Examples

1. 549: Testing for divisibility by 3. Sum of digits: 5+4+9=18. Since 18 is divisible by 3 hence 549 is divisible by 3.

2. 17497: 1+7+4+9+7= 28= 2+8=10. Since 10 is not divisible by 3, therefore 17497 is not divisible by 3

Divisible by 4:

Rule: A number is divisible by 4 if the last two digits form a number divisible by 4 or if the last two digits are 0.

Examples

1.7256. Since 56 is divisible by 4 therefore 7256 is also divisible by 4

2.8452800. There are two 0s at the end therefore the number is divisible by 4.

Note. The same rule is applicable to check for divisibility by 25. i.e. a number is divisible by 25 if it ends with 25 or 2 zeros.

Divisibility by 5:

Rule: A number is divisible by 5 if it ends with 0 or 5.

Example: 735, 1485, 21070, are all divisible by 5.

Divisibility by 6:

Rule: A number is divisible by 6 if the number is divisible by both 2 and 3.

Examples:

1. 7416, the last digit is 6. Test for divisibility by 2. 2 divides 6.

Test for divisibility by 3

Sum of digits 7+4+1+6=18. 3 divides 18. Therefore 7416 is divisible by 6

2. 5679173. The last digit 3 is not divisible by 2; therefore we can conclude that the number is not divisible by 6

Divisibility by 7

Rule: A number is divisible by 7 if the absolute difference between twice the unit digit and the number formed by the rest of the digits is 0 or divisible by 7 itself.

Examples

1. 343. The unit digit is 3. When you double 3 you have 3x2 =6. The number formed by the remaining digit is 34. Then 34-6=28. Since 7 divides 28, therefore 343 is divisible by 7.

Check the following to verify divisibility by 7

1. 644

2. 903

3. 1134

Divisibility by 8

Rule: A number is divisible by 8 if the last 3 digits form a number divisible by 8. Also if the last 3 digits are 0, the number is divisible by 8

Examples

1. 46328. Here the last 3 digits 328 is divisible by 8. Hence 46328 is divisible by 8

2. 4578000. The last 3 digits are 0 hence the number is divisible by 8

Note. This rule is also applicable to test for divisibility by 125

Divisibility by 9

Rule: A number is divisible by 9 if the sum of its digits is divisible by 9

Examples

1. 5148. Sum of digits 5+1+4+8=18. 18 is divisible by 9 hence 5148 is divisible by 9

2. 1085436. Sum of digits 1+0+8+5+4+3+6=27. 27 is divisible by 9 hence 1085436 is divisible by 9.

Divisibility by 10

Rule: A number is divisible by 10 if it ends with 0

530, 876540, 346790 are all divisible by 10

Divisibility by 11

Rule: a number is divisible by 11 if the sum of its digits in odd place minus sum of its digits in even places givers 0 or multiples of 11.

Examples

1. 53746275

Sum of digits in odd places=5+7+6+7=25

Sum of digits in even places=3+4+2+5=14

The difference 25-14=11

Since 11 divides 11, therefore 53746275 is divisible by 11

2. 3256682

Sum of digits in odd places=3+5+6+2=16

Sum of digits in even places=2+6+8+=16

The difference 16-16=0

Therefore 3256682 is divisible by 11

Divisibility by 12

Rule: A number is divisible by 12 if is divisible by both 3 and 4

Examples

To check quickly we first test for divisibility by 4. If that condition is satisfied we then proceed to check divisibility by 3.

Example

134892. To test divisibility by 4 we examine 92, since 4 divides 92 we proceed to test for 3.

Sum of digits 1+3+4+8+9+2=27. 3 divides 27

Therefore 134892 is divisible by 12 since it is divisible by both 3 and 4

Divisibility by 13

Rule: A number whose sum of four times the unit digit and the number formed by the rest of the digits is divisible by 13 is itself also divisible by 13. To check Add 4 times the last digit to the remaining number. Repeat the step as necessary. If the result is divisible by 13, the original number is also divisible by 13

Example

Show that 1248 is divisible by 13

The unit digit is 8 and 4x8=32

The remaining number is 124 and 124+32=156

We repeat the step for 156

The unit digit is 6 and 4x6=24

The remaining truncated number is 15 and 15+24=39 which is divisible by 13.

Hence 1248 is divisible by 13

Divisibility by 14

Rule: Any number which is divisible by both 2 and 7 is also divisible by 14 i.e. the last digit of the number is even and at the same time satisfies the condition for divisibility by 7.

Example

Show that 1344 is divisible by 14

Test for 2. 2 divides 4 which is the last digit hence 2 divides 1344

Test for 7. When we double the unit digit we have 2x4=8

134-8=126

When we repeat we have 12- (6x2) =0

Hence 7 divides 1344

Therefore 1344 is divisible by 14

Divisibility by 15

Rule: Any number which is divisible by both 3 and 5 is divisible by 15

Example

225

Test for 5. The last digit is 5 hence the condition for divisibility by 5 is satisfied

Next we test for 3. Sum of digits 2+2+5=9 hence the condition for divisibility by 3 is also satisfied.

Therefore 225 is divisible by 3

Divisibility by 16

Rule: Any number whose last four digits is divisible by 16 is also divisible by 16

Example

1578464 is divisible by 16 since 16 divides 8464

Divisibility by 17
Rule: When 5 times the last digit subtracted from the remaining truncated number gives 17 or a multiple of 17.

Then the number is divisible by 17. Repeat the step as necessary.

Example

Check 1683

The unit digit is 3 and 5×3=15

168-15=153

When the test is repeated

We have 15- (5×3) =0

Hence 1683 is divisible by 17

Divisibility by 18

Rule: Any number which is divisible by 9 and has its last digit even or 0 is divisible by 18

Example

926568

Test for divisibility by 9. Sum of digits 9+2+6+5+6+8=36 which is divisible by 9 and the unit digit "8" is even

Therefore 926568 is divisible by 18

Divisibility by 19

Rule: When you add 2 times the unit digit to the remaining number and if the result is divisible by 19, the original number is also divisible by 19. Repeat the step as necessary.

Example

Check if 1634 is divisible by 19

163+ (2x4) =171

When we repeat we have

17+ (2x1) =19

19 is divisible by 19 hence 1634 is divisible by 19

3. Shortcut Tricks for Multiplication and Division of Numbers

Deep understanding of relationship between multiplication and division will help in developing efficient personal strategies for mental manipulation of numbers. The good thing about computation in math is that a problem can be solved in a variety of ways depending on the level of understanding of the learner. The goal of this chapter is to teach you some easy yet impressive calculation you can learn to do immediately and move gradually to more complex problems.

The following exercises are useful guide for learners to preferred personal mental computation strategies.

Multiplication by powers of 10

We shall start with Multiplication by powers of 10

This is the foundation for deep understanding of number manipulation

To multiply a number with 10, count the number of 0 in the power of 10 and merge it to the whole number. Let us consider the table below.

X	n	x^n
10	0	1
10	1	10
10	2	100
10	3	1000

Examples

48x10

10 has only 1 zero digit. Merging this with 48 gives 480

48x10=480

48x100

100 has 2 zero digits. Merging t with 48 gives 4800

Thus

48x100=4800

Decimal numbers

When multiplying with a decimal, shift the decimal point to the right for each zero in the power of 10.

Examples

1. 0.87x10

    10 has only one zero digit. So we move the decimal point 1 place to the right. Thus 0.87x10 becomes 8.7
    0.87x10=8.7

2. 0.87x100

    100 has 2 zero digits, we shift the decimal point two places to the right.
    0.87x100=87

Negative Powers

When the exponent is a negative number, the decimal point one place to the left for each zero you see before the 1.

Remember that

$0.1 = 10^{-1}$

$0.01 = 10^{-2}$

$0.001 = 10^{-3}$

and so forth

Examples

87x0.1

There is only one zero, so we shift the decimal point one place to the left.

Thus

87x0.1=8.7

Example 2

87x0.01

Two zero digits before one. We shift the decimal point two places to the left.

0.87x0.01= 0.87

Extending the Idea of Multiplication by 10

You can use the powers of 10 even if this is not obvious in the given problem.

The trick is to break up one or both number to make it suitable for the desired operation then, multiply and combine the parts to make up the required result.

Examples

1. 25x7

Breaking up for suitability and quick result

25(4+3)

(25x4)+ (25x3)

100+75=175

2. 25x28

Breaking up for suitability

25x **(4x7)**

(25x4) x7

100x7=700

3. 75x24

25x3x4x6

25x4x3x6

100x18 =1800

Multiplying by a Number that is 1 less than a Power of 10. Such as 9, 99, 999, 9999.

Examples

Step 1

Round up 9 to the nearest power of 10

Step 2

Find the product of the power of 10 and the multiplier.

Step3

Subtract the original number from the result

Examples

1.58 x 9

Round up 9 to 10.

Product of 58 x 10 = 580

580 – 58 = 522

2. 58 x 99

58 x 100 = 5800

5800 − 58 = 5722

3. 58 x 999

58 x 1000 = 58000

58000 − 58 = 57942

Multiplication with Multiples of 5 that are Factors of Powers of 10

I.e. (**5, 25, 50, 125, 250, 500**)

The Short Cut Trick

The strategy below will help you carry out multiplication of these numbers within a short time.

Since all these numbers are factors of powers of 10. They can be written in another form to make them more suitable for manipulation.

Let us consider the following

5 can be written as 10/2

25 can be written as 100/4

50 can be written as 100/2

125 can be written as 1000/8

250 can be written as 1000/4

500 can be written as 1000/2

Step 1

Express the given number as a fraction with a denominator of powers of 10. (See the table above)

Step 2

Find the product of the given number with the result in step1 and simplify accordingly.

Examples

    1. 35 x 5

    2. 35 x 25

    3. 35 x 50

    4. 35 x 125

    5. 35 x 250

    6. 35 x 5000

Solution

1. 35 x 5

Use 10/2 for 5

Therefore 35x10 /2 =350/2=175

2. 35 x 25

Using 100/4 for 25 we have

35 x 100 /4= 3500/4= 875

3. 35 x 50

Using 100/2 for 50 we have.

35 x 100/2 = 3500 = 1750

4. 35 x 125

125 can be written as 1000/8

Therefore 35 x 1000/8 = 35000/8 = 4375

5. 35 x 250

250 can be written as 1000/4 = 35000/4 = 8750

6.35 x 500

500 can be written as 1000/2

35 x 1000/2 = 35000/2= 17500

Try the following exercises

1.65x5

2. 68x 125

3.432x500

4.23x250

5.185x125

6.483x11=5313

Multiplying Two Digit Numbers Ending with 1

When multiplying two digit numbers ending with 1.you can use the following methods to get the answer within a short time. Note when two numbers that has 1 in their last digit are multiplied the result also contain 1 in the last digit.

Step 1

Find the product of the digits in the tens place.

Step 2

Get the sum of these digits

Step 3

Place the result in step 2 next to the result from step 1.

Step 4

Put 1 next to the result from step 3.

Examples

1. 41x21   **2.**51x31   **3.**61x71   **4.**51x31

Solution

1.41x21

Step 1

The product of digits in the tens side

4 x 2 = 8

Step 2

The sum of 4 and 2

4 +2 = 6

Step 3

Putting the two results together we have 86.

Step 4

Placing 1 next to the result in step3 we have 861.

Thus 41 x 21 = 861

Or quickly

4x2=8

4+2=6

Placing 1 at the end gives 861

Example 2   51x 31 = 1581

Step 1

The product of 5 x 3 = 15

Step 2

The sum of 5 and 3 = 8

Step 3

Putting the result together we have 158

Step 4

Putting 1 at the end we have 1581

Thus 51x31=1581

Or quickly

5x3=15

5+3=8

Placing 1 gives 1581

Special cases

The strategy is modified slightly when the result of product and sum are more than 1 digit. This is better illustrated in the following examples.

Example 3

61x71

Step 1

The product of 6 x 7 = 42

Step 2

The sum of 6 and 7 is 13

Step 3

The putting the result together 4(2)(1)3. Note we write (42 13) as

(42+1) 3

433

Step 4

Put 1 next to the result in 3 we have 4331

Thus 61x71= 4331

Example 4

61 x 51

Step 1

The product if 6 and 5 = 30

Step 2

The sum of 6 and 5 = 11

Step 4

Merging the results 3(0)(1)1 (30 11) is written as (30+1) 1 which gives 311

31

Step 4

Putting 1 we have 3111

Multiplying Numbers Between 11 and 19

Step 1

Identify the larger number in the given number

Step 2

Add the number to the most digit of the second numbers

Step 3

Put 0 at the end of the result in step 2

Step 4

Multiply the right most digit of both numbers

Step 5

Add step 3 and step 4

Example 1

15 x 13

Step 1

15 is the highest

Adding this to the right most digit of 13

Step 2

15 + 3 = 18

Step 3

Putting 0 at the end 180

Step 4

Product of the right most digit 5 and 3 = 15

Step 5

Adding Step 3 and Step 4

180 + 15 = 195

Example 2

19 x 15

Step 1

19 is the highest

Step 2

Adding 19 and 5 gives 24

Step 3

Putting 0 to the result we have 240

Step 4

The product of 9 and 5 gives 45

Step 5

Adding Steps 3 and 4

240+45 = 285

Multiplying Two Numbers Ending with 5

It is easier to multiply two different numbers with 5 in their last digits. This is the procedure

Step1

Find the product of the ten digits

Step2

Find the average of this same number

Step3

Add the results in steps 1 and 2 and multiply this with 100

Step4

Add 25 to the result in step3 to get the final answer.

Examples 1.35x45   2.65x25

1. 35x45

    The digits in ten place are (3) and (4)

    Step1

    The product of 3 and 4 is **12**

    Step2

    The average of 3 and 4 is 3+4 =7\2 =**3.5**

    Step 3

    The sum of steps1 and 2 12+3.5=15.5x100= 1550

    Adding 25 we have 1550+ 25=1575

Example 2. 65x25

Step1

The product of 6 and 2 is 6x2=**12**

Step 2

The average of 6 and 2 is 6+2=8 and 8/2=**4**

Step 3

Adding steps1 and 2 gives 16. And 16x100=**1600**

Adding 25 we have 1600+25=**1625**

Verify the following results

45x55=2475

65x75=4875

85x25=2125

35x75=2625

55x95=5225

Multiplying Two Numbers with a Difference of 2.

Given any two numbers a and b with a-b = 2. The product can be calculated within a short period of time following the steps below:

Step 1

Find the mid number between the two numbers; alternatively add 1 to the least number.

Step 2

Square your result in step 1

Step 3

Subtract 1 from your result in step 2 to get the answer.

Examples

1. Find 24 x 26

2. Find 29 x 31

Solution

1. 24 x 26, since 26 − 24 = 2. Hence the method will work.

Step 1. 24 + 1 = 25

Step 2 squaring 25 $25^2 = 625$

Step 3 subtracting 1. 625 − 1 = **624**

Therefore, 24 x 26 = 624

2. 29 x 31. 31 − 29 = 2

Step 1. 29 + 1 = 30

Step 2. Squaring 30. $30^2 = 900$

Step 3. Deducting 1. 900 − 1 = 899

Therefore, 29 x 31 = **899**

Exercises

Verify the following results using the above trick

1. 19 x 21 = 399
2. 22 x 24 = 528
3. 59 x 61 = 3599
4. 91 x 89 = 8099
5. 101 x 99 = 9999

Multiplying two Numbers with a difference of 3

Given any two numbers a and b such that $a - b = 3$. The product of a and b can be obtained with this shortcut method.

Step 1

Add 1 to the least number

Step 2

Find the square of the result

Step 3

Subtract 1 from the same number in step 1 and add to the result in step 2 to get the final answer.

Examples

1. Find 43 x 46
2. 54 x 57

Solution

Step 1

The lowest number is 43. Adding 1 gives 43 + 1 = 44

Step 2.

Squaring the result in step 2. 44^2 = 1936

Step 3.

Subtracting 1 from the same number in step 1 gives 43 – 1 = 42. Adding this to the result we have 1936 + 42 = 1978.

Thus, 43 x 46 = 1978

Example 2

54 x 57

Step 1

Adding 1 to 54 gives 54 + 1 = 55.

Step 2

Squaring this 55^2 = 3025

Step 3

Subtracting 1 from the same number in step 1 we have

54 − 1 = 53 and adding gives 3025 + 53.

Thus, 54 x 57 = 3078

Multiplying Two Numbers with a Difference of 4.

When the difference between two numbers is 4, the short cut trick is discussed below.

Step1

Find the number midway between the two given numbers

Step 2

Square this number and subtract 4 from the result to get the final answer.

Examples

Find 1.28x32 and 2. 54x58

1. 28x32

Step1.

The mid number between 28 and 32 is 30

Step2

$30^2 = 900$

Step3.

900-4=896.

Thus 28x32=896

2.54x58

Step1

The mid number between 54 and 58 is 56

Step2

$56^2 = 3136$

Step3

3136-4=3132.

Thus 54x58=3132

Exercises

Verify the following results

1.22x26=572

2. 38x42=1596

3.47x51=2397

4. 64x68=4352

5. 95x97=9405

Multiplication of Two Digit Numbers

The strategy discuss in this section help to calculate the product of any two digit numbers within a short period of time. Remember constant practice makes it easier.

To find the product of two digit numbers and two digit number we follow these steps.

Step1

Set out the numbers in the usual way and multiply the unit digits of the number and enter your answer in the unit digit column of the answer.

Step2

Cross multiply the digits of both numbers i.e. the tens digit of the first number with the unit digit of the second number and vice versa and add the result with the earlier result in step 1

Step3

Multiply the numbers in the tens digit and enter it before the result in step 2

The examples below will make it easier to understand

1. 41x27   2. 52 x37

Solution

1. 41x27

Step 1 the product of 1 and 7 1x7=7

Enter the result under the unit digit in the normal way

41
<u>x27</u>
7

Step 2 cross multiplying the units and tens digits we have 4x7=28 and 2x1=2 and 28+2=30.

Enter 0 with the earlier result and carry 3

Write 0 and carry3

```
4       1
 ×
2       7
2+28=30
```

Step 3

The product of 4 and 2; 4x2=8 when you add the carryover from the earlier result we have 8+3=11. Entering this result at the appropriate column we have the final answer.

```
   4  1
   2  7
  ------
1  1  0  7
```

Example2   52x37

Step1

The product of 7 and 2; 7x2=14 enter 4 and carry 1

As usual we enter 4 and carry 1

Step2

Cross multiplying 5x7=35 and 2x3=6; adding 6+35=41+1 (the carryover) =42 enter 2 and carry over 4

Step 3

The product of 3 and 5 i.e. 3x5=15 adding 4 (the carryover) gives 19. Therefore we have 1924 as the final answer

52 x 37= 1224

Verify the following results

1.33x46=1518

2.54x67=3618

3.69x38=2622

4.84x64=5376

5.76x82= 6232

4. Shortcut Method for Multiplication of a Number by 11

Next we want to examine multiplication of a number by 11. Working with Multiplication by 11 is easy.

When multiplying 11 with any unit digit from 1-9, we double the digit to get the result. For example

2 x 11 = 22

3 x 11 = 33

4 x 11 = 44

9 x 11 = 99

However, when the number is greater than 9, we use a different trick.

Let us look at the following examples

    1. 35 x 11

    2. 27 x 11

    3. 53 x 11

Using calculator we obtain the following results

35 x 11 = 385

27 x 11 = 297

53 x 11 = 583

In addition, for better understanding let us consider the conventional method of multiplication as we are taught in the school.

```
1.    3 5          2.    2 7          3.    5 3
    x 1 1              x 1 1              x 1 1
      3 5                2 7                5 3
    3 5                2 7                5 3
    3 8 5              2 9 7              5 8 3
```

If you observe carefully, you will notice that:

1. The digit in the number multiply with 11 appears again in the answer.

2. Furthermore the number in the middle is just the addition of the two digits.

This is the basic skill needed when dealing with multiplication by 11 to get the result quickly. Ignore 11 at first and focus on the multiplier then **add** the two digits involved and insert the result in between the two numbers.

45 x 11

When you add 4 and 5 you get 9.

Put 9 in the middle of 4 and 5 to give **495**.

Thus 45 x 11 = 495

Or quickly

4 [4+5] 5 = 495

Similarly 52x11 gives 572 the addition of the two digits involved is 7. When this comes between 5 and 2 we have 572.

Or quickly

5 [5+2]   2= 572

Exercises

Verify the following results

1. 63x11=693

2. 72x11=792

3. 53x11=583

4. 54x11=594

5. 27x11=297

Special Cases

Note, when the sum of the multiplier is greater than 9 this result is not applicable directly, we need to make some adjustment to the trick as discussed below.

Let us look at 57 x 11. The sum of 5 and 7 is 12. Therefore 57 x 11 = 5 [5+7] 7 using our earlier argument. But 5x7 is not equal to 5 [12] 7

In this case when you add the two digits, your result will be in tens, 5 [12] 7 so write the remainder and add the carryover to the tens digit in the left digit of the given number.

In this example: 57 x 11. Split 12 by adding 1 to the left digit of the number and put the reminder '2' in the middle. Thus 57 x 11 = 627 effortlessly

57 x 11 = 5 [5+7] 7 = 5 [12] 7= 5 [10+2] 7 which gives 5+1 [2] 7 =627 when you carry 1 from the bracket and add to 5 in the left.

Other examples

47 x 11 = 4 [4 + 7] 7 = 4 [11] 7 = 4 [10 + 1] 7 which gives 4+1 [1] 7= 517 when you carry 1 from the bracket and add to 4 in the left

69 x 11 = 6 [6 + 9] 9 = 6 [15] 9 = 6 [10 + 5] 9 which gives 6+1 [5] 9= 759 when you carry 1 from the bracket and add to 6 in the left

88 x 11 = 8 [8 + 8] 8= 8 [16] 8 = 8 + 1 [10 + 6] 8 which gives 8+1 [6] 8= 968 when you carry 1 from the bracket and add to 8 in the left

When the multiplier is a 3 digit number

Consider the following.

1. 502x11 and

2. 512x11

3. 428x11

4. 519x11

```
    5   0   2              5   1   2
   x    1   1             x    1   1
    5   0   2              5   1   2
 +5  0   2             +5   1   2
   5   5   2   2          5   6   3   2


    4   2   8              5   1   9
  x     1   1            x      1   1
    4   2   8              5   1   9
 +4  2   8             +5   1       9
   4   7   0   8          5   7   0   9
```

Note: Again, the digits of the multiplier appear in the answer

The short cut trick

Step 1

Focus the number in the middle of the multiplier

Step 2

Add the number to the digit in the left and to the digit in the right. Merge the two numbers and

Step 3

Replace the middle number with the result to get the required answer.

Example 1.

512 x11

Step 1

The middle number is 1

Step 2

Adding with left and right digits we have

1+5 = 6

1 + 2 = 3

Merging gives 63

Thus 512 x 11 will give you 5**63**2 when you replace 1 with 63.

2.418 x 11

1+4 = 5

1 + 8 = 9

Merging 59

Thus 418 x 11 = 4**59**8

Special cases

In a situation when the sum of the middle digit and the left or right number is greater than 9 we adjust the trick slightly as in the following examples;

1. 572 x 11 = 6292

2. 678 x 11 = 7458

Solution

1. 572 x 11. Adding the middle digit to both left and right digits

7 +5= 12

7 + 2 = 9

When we merge we have (12)9 inserting this gives 5(12)97 which is not the value of 572x11. Therefore we will carry 1 and add to the left digit to obtain 6 and keep the remainder. Thus the middle number and the left digit of the multiplier now become 692. When you put 2 you have 6292. Thus 517x11= 6292

Or quickly

5 [(5+7) (7+2)]2

5 [(12) (9)] 2

= 6 [29]2 when you carry 1 and add to the left digit

2.678x 11.

Adding the middle digit to both left and right digits

7 + 6 = 13

7 +8 = 15

The two numbers (13) and (15) are greater than 9. First add 13 to the left digit which is done by keeping 3 and carry 1 thus the first digit becomes 6+1 R 3 which is recorded as 73. Then add the second result (15) with what you have. Which is done by keeping 5 and carrying 1 which gives 73+1 R 5 written as 745. Thus the middle number and the first digit become 745. When you merge this with 8 you have 7458. Thus 678 x 11 = 7458

Or quickly

678x11

6 [(7+6) (7+8)] 8

= 6 [(13) (15)] 8

When we carry 1 from 13 and add to the left digit we have

= 6+1 [(3) (15)] 8

We manipulate the middle term (3) (15) by carrying 1 from 15 and add to 3 to give 4

7 [(4) (5)] 8

Therefore 678x11=7458

Exercises

Use the trick to confirm the following results.

724 x 11=7964

823 x 11 =9053

471 x 11=5181

576 x 11=6336

5. Shortcut Strategies for Calculating Squares of Numbers

Table of Perfect Squares 1-30

$1^2=1$	$11^2=121$	$21^2=141$
$2^2=4$	$12^2=144$	$22^2=484$
$3^2=9$	$13^2=169$	$23^2=529$
$4^2=16$	$14^2=196$	$24^2=576$
$5^2=25$	$15^2=225$	$25^2=625$
$6^2=36$	$16^2=256$	$26^2=676$
$7^2=49$	$17^2=289$	$27^2=729$
$8^2=64$	$18^2=324$	$28^2=784$
$9^2=81$	$19^2=361$	$29^2=841$
$10^2=100$	$20^2=400$	$30^2=900$

Squaring Numbers Ending with 5 in 25 Seconds

We want to examine short cut tricks for squaring number within a short period of time

Consider the following

15x15=225

25x25=625

35x35=1225

If you observe the answer has a pattern of 25 in the last two digits. This is always the case when you are squaring two digits numbers ending with 5. The question now is how do we determine the other digits before 25. The trick is discussed below

To obtain the square within some few seconds

Step 1

Pick the digits in 10s place.

Step 2

Add 1 to the result in step 1

Step 3

Find the product of the two numbers in steps 1 and 2.

Step 4

Place the result on the left side of 25 to give you the square of the required number instantly.

Example 1

25^2

To find 25^2 i.e. 25 x 25

Step 1

2 is the digit in 10s place

Step 2

Adding 1 to 2 gives 3.

Step 3

The product of 2 and 3 is 6.

Step

Putting this before 25 we have 625.

Thus $25^2 = 625$

Or quickly

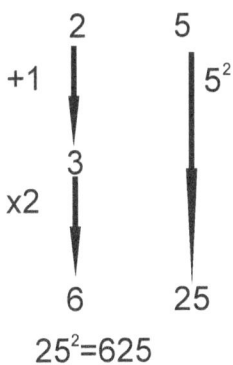

$25^2 = 625$

Example 2 45^2

Step1

The digit in tens place is 4.

Step2

Adding 1 to 4 gives 5.

Step3

The product of 4 and 5 is 20

Step4

When this is placed before 25

You have 2025.

Therefore $45^2 = 2025$

Or quickly

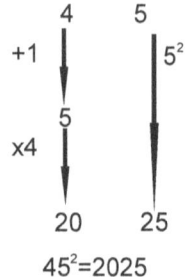

$45^2 = 2025$

Example 3   finds 65^2

65 x 65

Step1

6 is the digit in 10s place.

Step2

When 1 is added you get 7

Step3

The product of 6 and 7 = 42

Step4

When this is place before 25

Therefore: $65^2 = 4225$.

Squaring Number Ending with 25

Important fact

The last digit of the square of a number ending with 25 is **625**

Procedure

Step 1

Pick the digit before 25

Step 2

Square the number

Step 3

Divide the number from Step 1 by 2

Step 4

Add the result in Step 2 and Step 3

Step 5

Multiply the result in Step 4 by 10

Step 6

Write down 625 after the result in Step 5

Examples

Find the square of the following numbers

1.**325**   2. **425**

Step1

The digit before 25 is 3

Step2

Squaring 3 gives $3^2 = 9$

Step3

When 3 in step 2 is divided by 2 we have 3/2 = 1.5

Step 4

Adding the results in Step2 and Step3 9+1.5 = 10.5

Step5

Multiply the result by 10, 10.5 x 10 = 105

Step6

Putting this before 625, we have 105625

Therefore $325^2 = 105625$

Or quickly

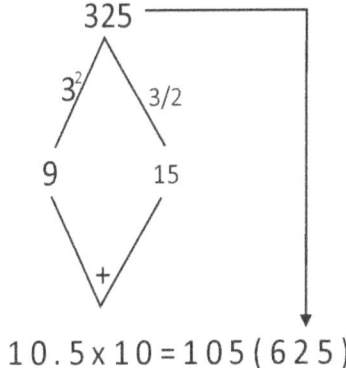

10.5 x 10 = 105 (6 2 5)

Example 2

425^2

Step1

The digit before 25 is 4

Step2

When 4 is squared we have $4^2 = 16$

Step 3

4/2 = 2

Step4

Adding the results in 2 and 3 we have 16+2=18

Step 5

Multiplying the result by 10 i.e. 18x10=180

Step 6

Placing the result before 625 gives 180625

Therefore

$425^2=180625$

Or quickly

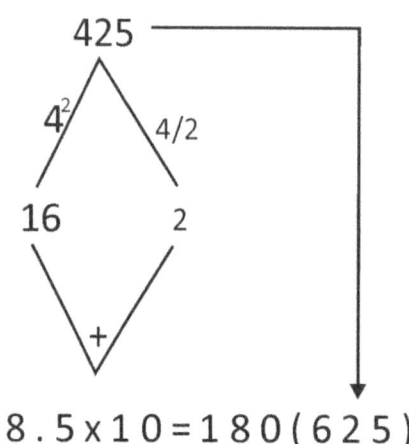

18.5 x 10 = 180(625)

Exercises

Verify the following results

1. $625^2 = 390625$

2. $825^2 = 680625$

3. $725^2 = 525625$

4. $875^2 = 765625$

5. $675^2 = 455625$

Squaring Number above 100 but less 130

Step1

Add the number in excess of 100 to the given value

Step2

Square the number and if the result is not up to two digits and put 0 before it

Step3

Place the result in step1 before 2

Examples

Square the following numbers

1. 102

2.103

Solution

1. 102^2

Step1

2 is the number in excess of 100, adding this to 102 gives

102 +2 =104

Step2

Squaring 2 we have 4

To make it two digits we place 0 to give 04

Step3

Merging the two result 10404

Thus $102^2 = 10404$

Or quickly

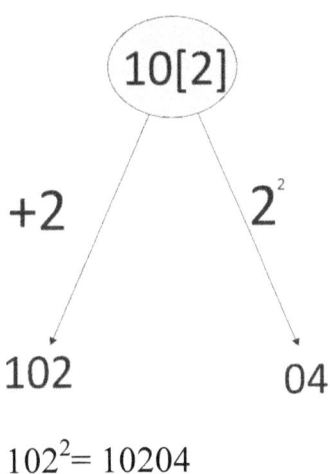

$102^2 = 10204$

2. 103^2

Step1

3 is the number in excess of 100 and adding this to 103 gives 106

Step2

Squaring 3 we have 09

Step3

Merging the result

10609

Therefore

$103^2 = 10609$

Or quickly

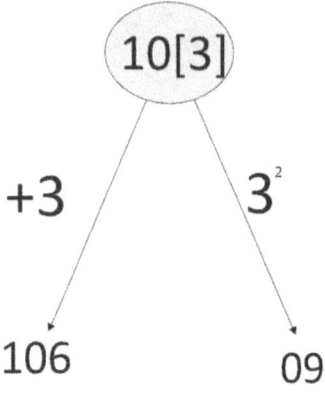

$103^2 = 10609$

When the number in excess of 100 is more than 9, twist step3 while merging by manipulating the digit in the middle

Examples

Find

1. 112^2

2. 116^2

Solution

1. 112^2 The number in excess of 100 is 12

Step 1

Add 12 to 112 gives 124

Step 2

Squaring 12 gives 144

Step 3

Merging the result 12(4)(1)44

Adding the middle digit

Therefore $112^2 = 12544$

Quickly

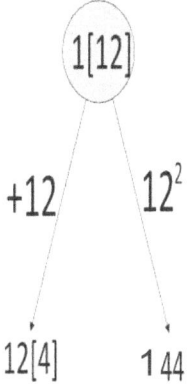

12(4+1)4=1254

112^2=1254

2. 116^2

The number in excess of 100 is 16

Step1

Add 16 to 116 gives132

Step 2

Squaring 16 gives 256

Step 3

Merging the result 13(2)(2)56

Adding the middle digit

Therefore 116^2=13456

Squaring a Number Below 100

This is the trick when we are to find the square of a number that is less than100

Step1

Subtract the number from 100

Step2

Subtract the result in step1 from the original number

Step3

Square the result obtained in step 1

Step4

Place the result in step3 next to result in step2

Examples find the squares of the following

1. 92

2. 89

Solution

1. 92^2

Step1

Subtract 92 from 100.   i.e.   100 -92 =8.

Step2

Subtract 8 from 92.   92-8 = 84.

Step3

Square the result obtained in step1 i.e. 8 = 64

Merging the result

Placing the result in step next to the result in step2

$92^2 = 8464$

Or quickly

100-92 = 8

92-8 8^2

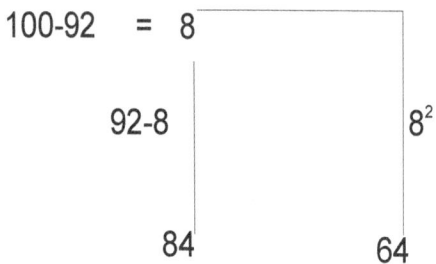

84   64

Merging 8464
Thus $92^2 = 8464$

2.89^2

Step1

Subtracting 89 from 100 gives 11

Step2

Subtracting 11 from 89 gives 78

Step3

Squaring 11

121

Merging the result

7(8)(1)21

=7(8+1)21

=7921

Quickly

100-89=11

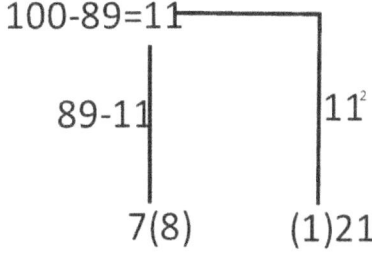

89-11 11^2

7(8)   (1)21

6. Shortcut Tricks for Square Roots of Numbers

X	x^2	Unit digit of x^2
1	1	1
2	4	4
3	9	9
4	16	6
5	25	5
6	36	6
7	49	9
8	64	4
9	81	1

Power table of squares

The tricks discuss in this section will enable you to find the square roots of any number within a short period of time. The condition is that the number must be a perfect square.

Please note these important points from the table above

1. If the last digit of a number is 1 the last digit of its square root is 1 or 9.

2. If the last digit of a number is 4 the last digit of its square root is 2 or 8.

3. If the last digit of a number is 5 the last digit of its square root must be 5

4. If the last digit of a number is 6 the last digit of its square root must be 4 or 6

5. If the last digit of a number is 9, the last digit of its square root must be 3 or 7

To find the square root of any number, the following steps will give the answer within a short period of time. Remember continuous practice will lead to speed and accuracy.

Step 1

First, determine the square root of the unit digit and partition the number into 2. Left hand side and Right hand side. (LHS & RHS). The first two digits in the number constitute the right hand side (RHS) and the remaining digit in the left side is LHS. For example, 841 is partitioned 8\ 41 with 8 in left hand side and 41 in the right hand side.

Step 2

Examine the left hand side and look for the next square number before and take the square root. In the above example our left hand side is 8 and the square immediately below it is 4, when you take the square root of 4 which is 2 as (answer 1). Take this as answer 1 and place it in the left digit of your final answer.

Step 3

To determine the unit digit

The last digit of 841 is 1. Which means the square root can be 1 or 9 from the table. To determine the actual digit we multiply (our *answer* 1) =2 with the next higher digit which is 3. Therefore, 2 x 3 = 6

Next we now compare it with the digit we have in the left hand side. Note if the calculated value is greater than the number in the left hand side. We will pick the smaller number (1 or 9) otherwise we pick the larger number as the unit digit of the problem. Our left hand side is 8 (841). The computed result is 6. Since 8 > 6 then we pick the highest of the two (1 and 9) which means the last digit of the square root of 841 is 9. Therefore, the square root of 841 is 29.

Examples

Find the square root of the following

1. 1296
2. 4489

1.**1296**

Note the last digit ends with 6 and the last digit of the square root must be 4 or 6.

Step 1

Partitioning the number 12|96

The left hand side 12 and the right hand side 96.

Step 2

12 is the number in the left hand side. The next higher square number before this is 9. Taking the square root we have square root 9 = 3. Therefore, we take 3 as the first

digit of our answer. As usual we call this answer one {Answer 1 = 3}

To determine the second digit we know that it can either be 4 or 6. The product of 3 and 4 is 12. Comparing this with left hand side gives 12 > 12

Since the calculated figure is not less than our left hand side therefore we pick the higher number. Consequently, the unit digit of square root of 1296 must be 6. Therefore, the square root of 1296 = 36

Example 2

4489

Solution

4489 ends with 9, therefore the unit digit of the square root must be 3 or 7.

Step 1

Partition the number. 44/89. Left hand side is 44 and right hand side is 89.

Step 2

Next higher square number before 44 is 36. The square root of 36 is 6. Therefore, 6 is the first digit of our answer. i.e. (answer 1 =6)

Step 3

To determine the second digit from the product of 6 and 7 we have 42 which is less than 44. Since the result is less than left hand side we pick the highest value (3 or 7) which is 7. Therefore the square root of 4489 is 67.

Exercises

Verify the following results

1. Square root of 961 = 31
2. Square root of 2809 = 53
3. Square root of 5476 = 74
4. Square root of 7396 = 86
5. Square root of 8649 = 93

7. Shortcut Strategies for Calculating Cube roots Of Numbers

The power table of cube

x	x^3	Unit digit
1	1	1
2	8	8
3	27	7
4	64	4
5	125	5
6	216	6
7	343	3
8	512	2
9	729	9

The shortcut trick discuss below will enable you to calculate the cube root of any number within a short period of time. For proper understanding of this trick

The number must be a perfect cube

- You must be able to identify cube of numbers from 1 to 9. (see the table)

Important points

The following points are obvious from the table.

1. If the last digit of a number is 0 the unit digit (last digit) of the cube must be 0.
2. If the last of a number is 1 the unit digit of the cube must be 1. $1^3 = 1$
3. If the last digit of a number is 2 the unit digit of its cube must be 8. $2^3 = 8$.
4. If the last digit of a number is 3 the unit digit of its cube must be 7. $3^3 = 27$
5. If the last digit of a number is 4 the unit digit of its cube must be 4. $4^3 = 64$
6. If the last digit of a number is 5 the unit digit of its cube must be 5. $5^3 = 125$
7. If the last digit of a number is 6 the unit digit of its cube must be 6. $6^3 = 216$.
8. If the last digit of a number is 7 the unit digit of its cube must be 3. $7^3 = 343$.

9. If the last digit of a number is 8 the unit digit of its cube must be 2. $8^3 = 512$.

10. If the last digit of a number is 9 the unit digit of its cube must be 9. $9^3 = 729$.

Cube roots

Cube root of any number can be calculated within a short period of time in just two steps.

Step 1

First determine the unit digit of the cube root of the number. See the table

Step 2

Ignore the last 3 digits of the given number and find the nearest perfect cube closer to the remaining digits.

The examples below give a good illustration

Find the cube roots of 12167 and 50653

Example 1

The cube root of 12167

Step 1

12167 has 7 as the last digit, therefore the unit digit of the cube root must be 3. (See the table)

Step 2

Ignoring the last three digits in 12167 we have 12.

The perfect cube before 12 is 8 which has a cube root of 2. Thus cube root 12167 = 23

Example 2

The cube root of 50653

Step 1

In 50653 the last digit is 3. Therefore, the unit digit of its cube must be 7.

Step 2

When we ignore the first 3 digits we have 50. The perfect cube closer to 50 is 27 with a cube root of 3. Hence, the cube roots of 50653 = 37.

Verify the following results

1. Cube root of 24389 = 29
2. Cube root of 19683 = 27

3. Cube root of 79507 = 43
4. Cube root of 175616 = 56
5. Cube root of 32768 = 32

8. Decimal Fractions

Fractions in which denominators are powers of 10 are known as decimal fractions.

Types of Fractions

Proper fractions: this is a fraction in which the numerator is less that the denominator. a/b such that a<b or b>a

Mixed fraction: this is an improper fraction in simplified form. 8/6 = 4/3 = 1 1/3

Important points

- Placing zeros after the decimal point at the extreme right of a decimal fraction does not affect the value of the number. For instance, 0.7 can be written as 0.70 or 0.7000
- If the numerator and the denominator of a fraction contain the same number of decimal places, we can safely remove the decimal points and evaluate the given problem. For example, 2.84/5.68 can be written as 284/568 = 1/2.

Addition and subtraction

Adding and subtracting decimals is done by arranging the given number in the right order using the place value system to make it suitable for the operation.

Example

Find $0.1 + 4.25 + 3.2 + 5 + 1.4253$

```
  0.1000
  4.2500
  3.2000
  5.0000
+ 1.4253
 13.9753
```

Multiplication of Decimal Fractions

To find the product of a decimal fraction within a short time take the following steps.

Step 1

Ignore the decimal points from the given problem and treat the number as a whole number.

Step 2

In the product obtained, place the decimal point after moving a number of places from the right equal to the sum of the decimal places in the given number.

Examples

Find

1. 0.03 x 0.4 x 0.52
2. 0.8 x 0.12 x 0.244

Solution

1. 0.03 x 0.4 x 0.52

Ignoring the decimal points

3 x 4 x 52 = 624

Note: your answer must be in 5 decimal places. Since the problem involved is 5 decimal places. The decimal point comes by adding two zeros to make 5 decimal places in the answer. Thus, 0.03 x 0.4 x 0.52 = 0.00624

2. 0.8 x 0.012 x 0.24

Multiplying ignoring decimal point we have 8 x 12 x 244 = 23424

Seven digits are involved after the decimal point.

Therefore, 0.8 x 0.012 x 0.244 = 0.0023424

Dividing a decimal fraction by a counting number

To divide a decimal fraction by a counting number.

Step 1

Ignore the decimal point to have set of whole numbers and divide.

Step 2

In the quotient obtained place the decimal point by starting at the right and moving a number of places equal to the sum of the decimal places in the given dividend.

Examples

1. 0.084 divided by 12.
2. 0.0306 divided by 17.

Solution

0.084 divided by 12.

Ignoring the decimal points we have 84 divided by 12 = 7. Since the dividend 0.084 is in 3 decimal places, the quotient must also be in 3 decimal places. Thus 0.084 divided by 12 = 0.007

0.0306 divided by 17.

Ignoring decimal point we have 306 divided by 17 = 18. Since the dividend 0.0306 is in 4 decimal places, the quotient must also be in 4 decimal places. Thus 0.0306 divided by 17 = 0.0018.

Dividing a decimal fraction by a decimal fraction

Decimal fraction

To divide a decimal fraction by a decimal fraction. Multiply both the dividend and the divisor by a suitable power of 10 to make the divisor a whole number and divide normally.

Examples

1. 0.0056 divided by 0.8
2. 0.00125 divided by 0.25

Solution

1. 0.0056/0.8

To make the divisor a whole number, multiply the numerator and the denominator by 10. Since it is in decimal place. 0.0056 x 10/ 0.8 x 10 = 0.056/8 = 0.007

2. 0.00125 divided by 0.25 = 0.00125/0.25

The denominator is in 2 decimal places therefore we have to multiply by 100.

0.00125 x 100/ 0.25 x 100 = 0.125/25 = 0.005

Converting of a decimal fraction to vulgar fraction

When converting a decimal to vulgar fraction

1. Change the denominator to a power of 10 then remove the decimal point and reduce the fraction to its lowest term.

Examples

Convert the following decimals to fraction

1. 0.4
2. 0.25
3. 2.008

Solution

1. 0.4 = 4/10 = 2/5
2. 0.25 = 25/100 = ¼
3. 2.008 = 2008/1000 = 502/250 = 251/125

Ordering of Fractions

A given set of fractions can be arranged in ascending or descending order within a short period of time. To compare and order a set of fractions, convert each fraction into decimals. When the numbers are in decimal it will be easier to arrange accordingly. Examples

Arrange the following set of fractions in ascending order

5/6, ¼, 3/5, 9/11 and ¾

Convert all to decimal

5/6 = 0.833

¼ = 0.25

3/5 = 0.6

9/11 = 0.818

¾ = 0.75

Arranging in ascending order

0.25<0.6<0.75<0.81<0.833

Thus, ¼, 2/5, 3/5, 9/11 and 5/6

9. Finding the Units Digits of Large Powers

Consider the series A,B,C,D,E,A,B,C,D,E,A,B,C,D,E,A,B,C,D,E,A,B,C,D,E.

We can find the nth term by counting up to n to get the term, for example if we are to find the 12th term from the series we can count starting from the first term A and stop at the 12th term which is B. Alternatively we can divide 12 by 5 (the series is a cycle of 5) each term is repeating itself after a cycle of 5. When 5 divides 12 it leaves a remainder of 2, which corresponds with letter B. Using this fact can be an easy way to calculate the last digit in a given problem.

The power table of 2,3,7 and 8

The power series of 2,3,7,8 share similar character as we shall see in the following tables.

Power table of 2

X	n	x^n	Last digit
2	1	2	2
2	2	4	4
2	3	8	8
2	4	16	6
2	5	32	2
2	6	64	4
2	7	128	8
2	8	256	6

Power table of 3

X	n	x^n	Last digit
3	1	3	3
3	2	9	9
3	3	27	7
3	4	81	1
3	5	243	3
3	6	729	9
3	7	2187	7
3	8	6561	1

Power table of 7

X	n	x^n	Last digit
7	1	7	7
7	2	49	9
7	3	343	3
7	4	2401	1
7	5	16807	7
7	6	117649	9
7	7	823543	3
7	8	5764801	1

Power table of 8

X	n	x^n	Last Digit
8	1	8	8
8	2	64	4
8	3	512	2
8	4	4096	6
8	5	32768	8
8	6	262144	4
8	7	2097152	2
8	8	16777216	6

These tables shows that the powers of 2,3,7. and 8 has a unit digit of a cycle of 4. The last digit repeats itself after a cycle of 4. This is an important fact in determining the unit digit whenever these numbers are raised to any power. The following examples will give further details.

Find the unit digits in the following questions

1. 2^{412}

2. 3^{358}

3. 7^{2848}

4. 8^{4567}

5. 642^{3872}

6. 387^{29545}

Solution

    1. 2^{412}.

From the table the unit digit of power of 2 is a cycle of 4 (2,4,8,6)

Therefore we divide the given power by 4, since 4 divides 412 perfectly, we can conclude that the unit digit will be 4th in the cycle of 4 (2,4,8,**6**). Hence the unit digit is 6

2. 3^{358}

The unit digit of the power of 3 has a cycle of 4 (3,9,7,1)

Therefore we divide the power by 4, when 4 divides 358 it leaves a remainder of 2. 357/4=39R2.

This means that the unit digit will be 2nd in the cycle of (3,**9**,7,1)

Hence the unit digit is 9.

3. 7^{2848}

From the table the unit digit of the power of 7 is a power of 4 (7,9,3,1). When 4 divides 2848 it leaves no remainder. Hence the unit digit must be 4th in the series (7,9,3,**1**)

Thus the unit digit 7^{2848} is 1

4. 8^{4567}.

The unit digit of the power of 8 is a cycle of 4 (8,4,2,6). When 4 divides 4567 it leaves a remainder of 3. Therefore the unit digit is the 3rd in the series (8,4,**2**,6).

The unit digit of 8^{4567} is 2

5. 642^{3872}

Recall 2 has a power cycle of 4.

When 4 divides the power it leaves a remainder of 3

Therefore the unit digit is the third of the cycle (**2**,4,8,6)

Which is 8

6. 387^{29545}

When 4 the divide the power it leaves a remainder of 1

Therefore the unit digit is first of the cycle(7,9,3,1)

Which is 7

Additional Facts

From the table, we can see that the cycle covers the last two digits in the power of 7. The last two digits also has a pattern of a cycle of 4 (07,49,43,01). This fact is useful in calculating the last two digits when 7 is raised to any positive integer within a short time.

Examples

Find the last two digits of 1. 7^{878} and 2. $7^{25434800}$

Solution

1. The last two digits of powers of 7 is a cycle of 4 (07,49,43,01). When 4 divide 878 it gives a remainder of 2. Ie 219R2. The last two digits is the 2nd of the cycle (07,**49**,43,01).

2. the last two digits of powers of 7 is a cycle of 4 (07,49,43,01). Since 4 divide 25434800 perfectly. Therefore the last two digits is the 4th in the cycle (07,49,43,**01**). The last two digits of $7^{25434800}$ is 01

The power tables of 4 and 9

The power tables of 4 and 9 share similar pattern

The power table of 4

X	n	x^n	Last Digit
4	1	4	4
4	2	16	6
4	3	64	4
4	4	256	6
4	5	1024	4
4	6	4096	6
4	7	16384	4
4	8	65536	6

Power Table of 9

X	n	x^n	Last Digit
9	1	9	9
9	2	81	1
9	3	729	9
9	4	6561	1
9	5	59049	9
9	6	531441	1
9	7	4782969	9
9	8	43046721	1

From these tables. The powers of 4 and 9 have a cycle of 2 pattern. The unit digit alternate between two numbers. In the case of 4 if the power is odd the unit digit is 4 and if the power digit is even the unit digit is 6. For 9 the unit digits alternate between 1 and 9 if the power is odd the unit digit is 9 and when it is even the unit digit is 1

Examples

Find the unit digit in the following questions

1. 4^{27}

2. 9^{48}

3. $4^{27} \times 9^{48}$

4. $189^{4345893}$

5. $1239^{38554248}$

6. 254^{39956}

7. 234^{108947}

8. 934^{400876}

Solution

    1. 4^{27}

The power (27) is odd hence the unit digit is 4. Recall the last digit in powers of 4 can either be 4 or 6

2. 9^{48}

The power (48) is even hence the unit digit is 1. The last digit in powers of 9 can either be 1 or 9

3. $4^{27} \times 9^{48}$

The unit digit in 4^{27} is 4

The unit digit in 9^{48} is 1

Therefore the unit digit in $4^{27} \times 9^{48}$ is 4x1=4

4. $189^{4345893}$

The power is odd

Therefore the unit digit is 9

5. $1239^{38554248}$

The power is even

Therefore the unit digit is 1

6. 254^{39956}

The power is even

Therefore the unit digit is 6

7.234^{108947}

The power is even

Therefore the unit digit is 4

8.934^{400876}

The power is even

Therefore the unit digit is 6.

Note

These facts can be recalled with this simple technique

$4^{odd}=4$ and $4^{even}=6$; similarly $9^{odd}=9$ and $9^{even}=1$

Just examine the last digit of the power (rule for divisibility by 2) and draw your conclusion if it is even or odd.

The Power Tables of 5 and 6

The power table of 5

X	n	x^n	Last Digit
5	1	5	5
5	2	25	5
5	3	125	5
5	4	625	5
5	5	3125	5
5	6	15625	5
5	7	78125	5
5	8	390625	5

The power table of 6

X	n	x^n	Last Digit
6	1	6	6
6	2	36	6
6	3	216	6
6	4	1296	6
6	5	7776	6
6	6	46656	6
6	7	279936	6
6	8	1679616	6

The powers of 5 and 6 have the same characters in that it is only the same number repeating itself.

The unit digit in any power of 5 is 5 and also in any power of 6 it is 6.

Other examples

Find the unit digits in the following

1. 5^{283} 2. 5^{16490} 3. 6^{215} 4. 6^{3780}

5. 315^{285423} 6. $946^{3194652}$

Solution

1. The answer is 5

2. The answer is 5

3. The answer is 6

4. The answer is 6

5. The answer is 5

6. The answer is 6

Summary

Memorizing the power cycle of all the digit from 1-10 will be of great help and promote fluency in the exercise

Number	Cyclicity	Power cycle
1	1	1
2	4	2,4,8,6
3	4	3,9,7,1
4	2	4,6
5	1	5
6	1	6
7	4	7,9,3,1
8	4	8,4,2,6
9	2	9,1
10	1	0

About the Author

David Itanola has taught mathematics at various levels for more than 25 years. This book is the result of his extensive research on different techniques and concepts on quick math. Apart from writing on mathematical subjects, David has a penchant for helping people living a productive life.

David holds the master's degree in Mathematics. He is married with three children and enjoys playing the piano and mentoring children.

www.ingramcontent.com/pod-product-compliance
Lightning Source LLC
Chambersburg PA
CBHW070151230526
45471CB00002B/617